For my two incredible reasons why. Without whom, I would not have the stories to tell about our fun adventures with our little herd of mini moos. You've pushed me to be the very best version of myself.

-Love Mama

Big mama lay down in the field soaking up the sun.

One day soon, she'll be watching her little one run.

Off in the distance, she hears the bucket shake.

That's a sound she knows well, make no mistake.

Here comes the little boy she's come to love so well.

He was bringing something nice and sweet,
she could already tell.

Biggg stretch up, have to make it to the gate.

Little boy moans and groans, the bucket had some weight.

He gives her a big scoop of cubes so she thanks him with a wet kiss on his face.

Then she felt his tiny hands hold her in his embrace.

She thought to herself right then and there "he'll love my calf as much as me".

For instead of just the two of them, it would now be three.

www.ingramcontent.com/pod-product-compliance
Lightning Source LLC
Chambersburg PA
CBRC102011060526
44107CB00156B/1265